THE POETRY OF ARGON

The Poetry of Argon

Walter the Educator™

SKB

Silent King Books a WhichHead Imprint

Copyright © 2023 by Walter the Educator™

All rights reserved. No part of this book may be reproduced in any manner whatsoever without written permission except in the case of brief quotations embodied in critical articles and reviews.

First Printing, 2023

Disclaimer
This book is a literary work; poems are not about specific persons, locations, situations, and/or circumstances unless mentioned in a historical context. This book is for entertainment and informational purposes only. The author and publisher offer this information without warranties expressed or implied. No matter the grounds, neither the author nor the publisher will be accountable for any losses, injuries, or other damages caused by the reader's use of this book. The use of this book acknowledges an understanding and acceptance of this disclaimer.

> "Earning a degree in chemistry changed my life!"
> – Walter the Educator

dedicated to all the chemistry lovers, like myself, across the world

CONTENTS

Dedication v

Why I Created This Book? 1

One - The Enigma 2

Two - Subtle Presence 4

Three - Peaceful Star 5

Four - Tranquil Spirit 7

Five - Treasure Divine 9

Six - Invisible Beauty 10

Seven - Quiet Magnificence 12

Eight - Infinite Grace 14

Nine - Noble Gas 16

Ten - Celestial Meditation 18

Eleven - Cosmic Harmony 20

Twelve - Boundless Sky 22

Thirteen - Tranquil Emblem 23

Fourteen - Symbol Of Peace 24

Fifteen - Argon's Elegance 25

Sixteen - Majestic And Calm 27

Seventeen - Dispelling Cosmic Scars 29

Eighteen - Stable And Resolute 31

Nineteen - Peace Without Flaw 33

Twenty - Secrets Of The Night 35

Twenty-One - Twirl And Spin 37

Twenty-Two - Soothing Balm 39

Twenty-Three - Inert Companion 41

Twenty-Four - Cosmic Ballet 42

Twenty-Five - Silent Repose 44

Twenty-Six - Celestial Decree 46

Twenty-Seven - Night And Day 48

Twenty-Eight - Noble Silence 50

Twenty-Nine - Essence Of Love 51

Thirty - Vast Unknown 53

Thirty-One - Divine Finesse 55

Thirty-Two - O Noble Argon 57

Thirty-Three - Dance Of Atoms	. . . .	59
Thirty-Four - Collide And Galaxies	. . . . . .	61
Thirty-Five - Through And Through	. . . . . .	63
Thirty-Six - Unyielding	. . . .	65
Thirty-Seven - Stillness	. . . .	66
Thirty-Eight - Argon, The Tranquil	. . . . . .	68
Thirty-Nine - Argon's Embrace	. . . .	70
About The Author	. . . .	72

WHY I CREATED THIS BOOK?

Creating a poetry book about the chemical element Argon was a unique and intriguing endeavor. Argon, a noble gas, is often overlooked in favor of more reactive elements. Crafting poetry about Argon could bring attention to its often unnoticed presence in our world, exploring its inert nature and symbolic significance. This creative exploration can offer a fresh perspective on an element that is often taken for granted.

ONE

THE ENIGMA

In the quiet realm of noble gases, Argon reigns supreme,
A silent sentinel in the air, its presence seldom seen.
Unmoving, unyielding, it holds its stoic place,
A noble element, with an enigmatic grace.
   Argon, the ghostly wanderer, in the atmosphere it dwells,
A solitary figure, where no story it tells.
Unreactive, unassuming, it watches the world go by,
Inert and unobtrusive, beneath the endless sky.
   In the glow of electric light, it casts a gentle hue,
An ethereal luminescence, a subtle, ghostly view.
In the depths of the Earth, it slumbers in the stone,
A quiet, hidden essence, in a world it calls its own.
   Argon, the enigma, in its solitude it thrives,
Unnoticed and unspoken, where the stillness thrives.

A mystery in its essence, a secret never told,
In the tapestry of elements, a story yet untold.

TWO

SUBTLE PRESENCE

In the quiet expanse of the atmosphere high,
Lies an element noble, without a need to vie.
Argon, the silent observer, unreactive and still,
Existing in solitude, with an unassuming will.
A gentle glow it emits in the electric light's embrace,
A subtle presence within the Earth, a mystery in space.
Unnoticed, unassuming, yet steadfast in its place,
Argon, the enigma, with elegance and grace.
In the tapestry of elements, it remains a hidden thread,
Thriving in solitude, where silence is widespread.
Oh, Argon, noble gas, so peaceful and serene,
In your unassuming nature, a beauty to be seen.

THREE

PEACEFUL STAR

In the quiet heart of the periodic table,
Lies a noble gas, tranquil and stable,
Argon, the silent observer in the night,
Emitting a gentle glow, a soft, soothing light.
Unreactive, still, and unassuming,
It exists in solitude, gracefully blooming,
A mysterious enigma, thriving in peace,
Adding a hidden thread to the elements' tapestry.
In the depths of Earth, it quietly resides,
A peaceful presence, where it confides,
In the vastness of space, its mystery unfurls,
A serene enigma, among the cosmic swirls.
Argon, the elegant dancer of the elements' stage,
A quiet performer, yet wise and sage,
In the presence of electric light, it comes alive,
A shimmering beauty, hidden in plain sight.

Oh, Argon, with your enigmatic grace,
You weave through the cosmos, leaving no trace,
A silent witness to the universe's grand design,
In your solitude, a peaceful star you shine.

FOUR

TRANQUIL SPIRIT

In the dance of elements, a serene enigma glows,
Argon, the peaceful presence, gracefully flows.
Amidst the chaos, it holds its tranquil stance,
A shimmering beauty in the electric light's dance.

Unseen, yet ever-present, in the air we breathe,
A noble gas, in solitude, it does not seethe.
Stable and unreactive, it quietly abides,
In the cosmic expanse, where mystery resides.

In noble stillness, it fills the starry night,
A celestial companion, shining ever so bright.
In Earth's embrace, it lingers without a trace,
An enigmatic essence, in its peaceful grace.

Like an elegant dancer on the stage of creation,
Argon whispers its tale, a silent revelation.
In its gentle glow, a beauty so sublime,
A tranquil spirit, frozen in the sands of time.

So let us marvel at this element's art,
A masterpiece of nature, a work of quiet heart.
For in its peaceful presence, we find our delight,
Argon, the noble gas, a spectacle so right.

FIVE

TREASURE DIVINE

In the stillness of the night, Argon silently glows,
A tranquil presence, a hidden thread only few knows.
Unreactive and serene, it dances with electric light,
A peaceful enigma, shimmering in the night.

Amidst the chaos, it flows with grace,
A celestial companion, in its tranquil space.
Unseen, yet ever-present in the air we breathe,
Stable and steadfast, it doesn't deceive.

A masterpiece of nature, elegant and rare,
In solitude, it exists, beyond compare.
With gentle radiance, it lights up the dark,
An unsung hero, leaving its mark.

Oh, Argon, you are a marvel to behold,
Your tranquil spirit, a story untold.
In the tapestry of elements, you quietly shine,
A symbol of stability, a treasure divine.

SIX

INVISIBLE BEAUTY

In the still of night, a glow so rare,
Argon, noble gas, beyond compare.
Unseen, untamed, in the atmosphere,
Its tranquil presence, oh so clear.

A ghostly light, a whispering hush,
Argon, in darkness, a gentle blush.
Mysterious element, serene and pure,
In the cover of night, it does endure.

With quiet strength, it holds its ground,
Inert and peaceful, a calm profound.
A dancer in twilight, a silent ballet,
Argon's allure, it never does sway.

Amidst the stars, it quietly gleams,
Invisible beauty, beyond our dreams.
A guardian of peace, a celestial sight,
Argon, in darkness, shines so bright.

So, let us cherish this enigmatic gas,
Invisible beauty, as the moments pass.
For in the still of night, it softly glows,
Argon, a wonder that nature bestows.

SEVEN

QUIET MAGNIFICENCE

In the stillness of the night, Argon's gentle glow,
A silent sentinel in the celestial show.
Unseen, yet steadfast, in the cosmic expanse,
A noble presence, in the dance of chance.

Unassuming, yet radiant, in its tranquil embrace,
Argon whispers secrets, through time and space.
A lustrous gem, in the tapestry of elements,
Elegance personified, in its silent settlements.

Unfazed by the chaos, of the world around,
Argon's serene countenance, knows no bound.
A shimmering specter, in the depths of the unknown,
A celestial wanderer, in the starlight it's sown.

In the depths of the Earth, and the boundless sky,
Argon's enigmatic allure, will never die.

A symphony of atoms, in the grand design,
Argon shines brightly, in the cosmic shrine.
  So, let us marvel at this noble gas so rare,
In its quiet magnificence, beyond compare.
For in the grand narrative of the universe's tale,
Argon's elegance and grace shall forever prevail.

EIGHT

INFINITE GRACE

In the quiet depths of the cosmos, Argon lies,
A serene presence beneath the starry skies,
Unassuming and tranquil, it silently drifts,
A noble gas, in the universe it shifts.

Amidst the celestial dance, it holds its own,
A noble solitude, in the cosmic zone,
With noble elegance, it calmly abides,
In the cosmic tapestry, where it quietly hides.

In the heart of distant galaxies, it lingers,
A mystic essence, with enigmatic fingers,
In the void of space, it gracefully glows,
A peaceful aura, that the universe bestows.

As the constellations weave their timeless tale,
Argon watches, a silent witness, so pale,
In its timeless gaze, a wisdom profound,
A celestial mystery, in the universe found.

So let us honor this noble gas so rare,
In its quiet presence, a cosmic affair,
For in its calm embrace, we find our place,
Amidst the stars, in infinite grace.

NINE

NOBLE GAS

In the tranquil embrace of the cosmic night,
Argon, noble and serene, dances with delight.
A shimmering essence, so gracefully bright,
In the quiet expanse, a celestial sight.

Unyielding stability, a steadfast friend,
Invisible yet present, till the universe's end.
Elegantly aloof, in its tranquil blend,
A cosmic companion, on which we depend.

With a quiet elegance, it fills the void,
A beacon of light, in darkness deployed.
In noble silence, it is ever employed,
A celestial whisper, never to be destroyed.

Oh, Argon, enigmatic essence of the sky,
In your peaceful presence, we can't help but sigh.
Shining brightly in the cosmic lullaby,
A timeless wisdom, soaring high and nigh.

Let's cherish and honor this noble gas,
For its quiet magnificence, and timelessness surpass.
In the grand design of the cosmic mass,
Argon's significance, let's embrace and amass.

TEN

CELESTIAL MEDITATION

Amidst the celestial dance, you gracefully reside,
Argon, noble and serene, in the universe wide.
Silent guardian of peace, in the cosmic expanse,
Your tranquil presence, a celestial trance.

Elegant whispers in the darkness, you softly glow,
A radiant luminary, in the cosmic show.
Enigmatic and noble, a celestial companion,
In the tapestry of stars, you shine as a champion.

A witness to the universe's tale, you quietly gleam,
Reflecting the cosmic symphony, like a tranquil dream.
Stoic and steadfast, a symbol of stability,
In the cosmic ballet, you embody tranquility.

In your noble silence, secrets find their keep,
A celestial enigma, in the vast cosmic sweep.

In your quiet elegance, a cosmic revelation,
Argon, you are the essence of celestial meditation.

ELEVEN

COSMIC HARMONY

In the quiet vault of the universe, there lies
An element serene, a tranquil prize
Argon, noble gas, in stillness it dwells
A celestial companion, its enigma compels
    Amidst the cosmic dance, it holds its place
With grace and elegance, it fills the space
Unseen, yet shining with a soft, steady light
A beacon of peace in the vast expanse of night
    In noble silence, it weaves its mystic lore
A guardian of secrets, it holds much more
Than meets the eye, in its calm embrace
Whispers of serenity, in every cosmic space
    Oh Argon, celestial sentinel so fair
Your presence brings solace, beyond compare
In the celestial ballet, you quietly gleam
A symbol of peace, in the cosmic dream

So let us cherish this enigmatic guest
In its tranquil aura, let our hearts find rest
For in the boundless expanse, it's plain to see
Argon, the essence of cosmic harmony

TWELVE

BOUNDLESS SKY

In the cosmic tapestry, a noble presence reigns,
Elegant and serene, Argon calmly abides,
A silent sentinel in the boundless expanse,
Unyielding, enigmatic, where mystery resides.

Deep within the Earth, it whispers its name,
In the hidden depths, where secrets lay,
A quiet witness to the ancient dance of time,
In noble solitude, it holds its sway.

Boundless sky, where stars ignite the night,
Argon's enigmatic allure, a shimmering light,
A beacon of peace, in the cosmic expanse,
A symbol of serenity, in the celestial dance.

In noble solitude, it claims its rightful place,
A silent witness to the wonders of space,
A tranquil force, in the grand design,
Argon, the noble element, so divine.

THIRTEEN

TRANQUIL EMBLEM

In the cosmic expanse, a noble presence reigns,
Argon, serene and enigmatic, it sustains.
Unseen, yet ever-present, in the celestial dance,
Elegance and grace, in its tranquil trance.

A beacon of peace, amid the cosmic storm,
Argon, the celestial companion, so warm.
Stability it brings, to the universe wide,
In its peaceful embrace, secrets do confide.

A silent sentinel, in the cosmic expanse,
Argon, symbol of peace, in its gentle stance.
Guardian of mysteries, it keeps its vigil true,
In its noble aura, serenity imbues.

So let us marvel at Argon's mystic allure,
Embracing its enigma, tranquil and pure.
In the cosmic tapestry, a celestial gem,
Argon, eternal, in its tranquil emblem.

FOURTEEN

SYMBOL OF PEACE

In the cosmic expanse, a noble solitude,
Argon, tranquil and serene, in quietude.
Unseen, yet steadfast in its peaceful aura,
A cosmic companion, with grace and allure.

Amidst the stars, it dances in timeless flight,
A celestial whisper, a beacon of light.
In noble silence, it holds its tranquil sway,
A timeless witness in the grand display.

With elegance and poise, it fills the night,
A symbol of peace, a source of quiet delight.
In the tapestry of the universe, it weaves,
A steadfast presence, in which harmony cleaves.

So let us marvel at this enigmatic gas,
A cosmic artist, painting serenity en masse.
In its embrace, find solace and repose,
For Argon, in its stillness, forever glows.

FIFTEEN

ARGON'S ELEGANCE

In the cosmic dance, a silent sentinel stands,
Argon, noble gas, where tranquility expands.
Amidst the stars, it holds its serene sway,
A beacon of peace in the celestial array.

With regal grace, it fills the boundless night,
A cosmic companion, gleaming pure and bright.
In its tranquil embrace, chaos finds repose,
As Argon's elegance through the cosmos flows.

Unseen, yet felt, in the depths of space,
It weaves a tapestry of serenity and grace.
A celestial waltz, where silence takes the lead,
And Argon's noble essence meets every need.

In the quiet expanse, it whispers a tale,
Of majestic calm and tranquility's prevail.
Oh, Argon, enigmatic and steadfast in flight,
You paint the universe with your serene light.

So let us honor this element so rare,
For in its noble silence, peace beyond compare.
Argon, the cosmic poet, in stillness profound,
In every corner of the cosmos, your presence is found.

SIXTEEN

MAJESTIC AND CALM

In the celestial expanse, Argon reigns,
A silent sentinel, its presence sustains.
Guardian of secrets, enigma profound,
In noble elegance, tranquility is found.

Amidst the stars, it quietly gleams,
A beacon of peace, in cosmic streams.
Witness to wonders, in silence it stays,
Embracing the universe in tranquil displays.

A noble gas, with grace it adorns,
In its serene embrace, harmony is born.
Majestic and calm, it paints the night sky,
Argon, the celestial, in peaceful supply.

In the tapestry of space, it weaves its thread,
A symbol of calmness, where chaos is shed.
Eternal and poised, it whispers its song,
Argon, the essence of tranquility, strong.

So in the cosmic dance, let its presence be known,
Argon, the noble, on its peaceful throne.
A celestial ode to its tranquil allure,
In the cosmic symphony, forever pure.

SEVENTEEN

DISPELLING COSMIC SCARS

In the cosmic expanse, a silent sentinel reigns,
Argon, in tranquil majesty, forever sustains.
Amidst the celestial dance, it holds its serene sway,
A timeless witness to the stars' luminous display.

Unseen and enigmatic, it glows with mystic allure,
A cosmic companion, steadfast and pure.
Its essence whispers of peace in the cosmic night,
A celestial gem, aglow with tranquil light.

In the embrace of Argon, chaos finds repose,
As it weaves through the cosmos, a tranquil prose.
Elegant and graceful, it dances with the stars,
Bringing serenity and harmony, dispelling cosmic scars.

Steadfast and stable in the universe's grand design,
Argon, a cosmic poet, in silence does entwine.

A witness to wonders, a source of eternal peace,
In the vast expanse of the universe, its tranquility will never cease.

EIGHTEEN

STABLE AND RESOLUTE

In the quiet depths of cosmic night,
Where stars ignite their radiant light,
There dwells a noble and tranquil soul,
Argon, a presence that makes the cosmos whole.
 Unseen, yet ever steadfast and true,
A timeless witness to the celestial view,
In elegance and grace, it silently glides,
Dispelling chaos where its essence resides.
 Through the boundless expanse, it weaves,
A tranquil prose, a cosmic peace it cleaves,
A poet of serenity, a harbinger of calm,
In its embrace, the universe finds a soothing balm.
 Stable and resolute, it holds its sway,
Amidst the dance of stars in astral array,

A companion enigmatic, yet ever bright,
Bringing harmony to the celestial night.
    O Argon, your essence so pure and serene,
A beacon of peace in the cosmic scene,
In your embrace, chaos finds release,
As you adorn the heavens with timeless peace.

NINETEEN

PEACE WITHOUT FLAW

In the cosmic dance, a silent witness stands,
Argon, tranquil, amidst the celestial expanse.
A poet of serenity, it weaves its tale,
Bringing elegance and grace, never to fail.

Amidst the chaos, it holds its ground,
Dispelling tumult with a resolute sound.
Where stars collide and galaxies swirl,
Argon's stability, a tranquil unfurl.

Eternal peace, its gentle embrace,
A beacon of harmony in the cosmic space.
Unyielding, unwavering, it remains,
A timeless essence, devoid of chains.

In its noble silence, it speaks volumes,
A symphony of calm, where chaos consumes.

Majestic and serene, its presence sublime,
Argon, the tranquil poet of space and time.
    A noble element, noble and true,
In its stillness, the universe finds its cue.
So let us behold, in wonder and awe,
Argon, the essence of peace without flaw.

TWENTY

SECRETS OF THE NIGHT

In the tranquil embrace of Argon's glow,
A cosmic dance of serenity and peace,
It weaves through the stars with grace untold,
A timeless witness to the universe's masterpiece.

Amidst the chaos of celestial ballet,
Argon stands as a beacon of tranquility,
A gentle guardian of the cosmic array,
Bringing harmony to the celestial city.

Its elegant essence, a shimmering light,
Unveiling the secrets of the night,
In its gentle embrace, chaos takes flight,
As Argon whispers, "All is right."

A cosmic poet, weaving verses of calm,
In the celestial symphony, a soothing balm,

Eternal and steadfast, a tranquil qualm,
Argon, the essence of cosmic psalm.
    So let us gaze upon its shimmering hue,
And find solace in its celestial view,
For in Argon's embrace, we find anew,
The peace that the universe bestows on few.

TWENTY-ONE

TWIRL AND SPIN

In the celestial dance, you quietly reside,
Argon, tranquil guardian of the starlit sky.
Amidst the chaos of cosmic storms that collide,
Your serene embrace brings peace, never to defy.

Inert and noble, you watch the universe unfold,
A witness to the wonders, a sentinel untold.
Your calm presence a balm to the celestial brawl,
Dispelling discord, bringing harmony to all.

In the shimmering nebulae, where chaos reigns,
You shimmer with a quiet, subdued grace,
A beacon of stability amid celestial chains,
In your tranquil aura, tumult finds no place.

In the vast expanse, where galaxies twirl and spin,
You hold the secrets of eons within.
Argon, element of peace, in your silent reign,
You soothe the cosmos, a serenity to sustain.

So let the stars weave their tales of grandeur,
For you, Argon, are the quiet, steadfast observer.
In your noble stillness, chaos finds its release,
As you eternally guard the universe in peace.

TWENTY-TWO

SOOTHING BALM

In the depths of cosmic night, a silent sentinel reigns,
Argon, noble and serene, in its celestial domains.
A ghostly glow, mysterious and profound,
It whispers ancient tales without making a sound.

Amidst the stars' luminous display, Argon stands tall,
A timeless witness to the cosmic dance, never to fall.
Unseen and enigmatic, yet its presence is felt,
As it weaves through the universe, where mysteries dwelt.

Glowing with mystic allure, it dances in the night,
A companion to the heavens, an enigma of light.
In the cosmic symphony, it plays its silent part,
Guiding the constellations with its enigmatic art.

Argon, a cosmic poet, brings serenity and grace,

In its tranquil embrace, the universe finds its place.
Elegance personified in its serene, steady gaze,
It paints the cosmic canvas with its gentle, mystic blaze.

A beacon of peace amidst the chaos of the cosmos,
Stability and resolute nature, in its tranquil repose.
A guardian of harmony, a keeper of cosmic calm,
Argon, the gentle sentinel, brings peace like a soothing balm.

TWENTY-THREE

INERT COMPANION

In the cosmic dance, Argon quietly abides,
A tranquil sentinel, where chaos subsides.
Unmoved by the tumult, it holds its serene sway,
In the celestial tapestry, it softly holds sway.

An inert companion to the stars' luminous display,
A witness to the heavens, in its own gentle way.
Unruffled by the fiery passion of the celestial fight,
It brings a sense of calm to the tumultuous night.

Amidst the cosmic storms and the astral parade,
Argon's quiet presence brings a soothing cascade.
A beacon of peace in the celestial expanse,
Dispelling the chaos with its tranquil advance.

In the vastness of space, where turmoil may reign,
Argon brings harmony, a peaceful refrain.
A celestial symphony, where tranquility thrives,
In the cosmic embrace, where Argon quietly survives.

TWENTY-FOUR

COSMIC BALLET

In the quiet realm where stars dance free,
Lies Argon, a tranquil symphony.
A timeless witness to the cosmic play,
It weaves serenity in the Milky Way.

Amidst the chaos of the celestial city,
Argon stands tall, a beacon of tranquility.
A poet of peace in the cosmic ballet,
Dispelling disorder with its gentle sway.

In the embrace of night, it softly glows,
Guardian of harmony, as the universe flows.
A noble gas, noble in its grace,
Embracing the cosmos in a tranquil embrace.

Majestic in its stillness, a calming force,
In the cosmic tapestry, it charts its course.
With quiet strength, it holds chaos at bay,
Argon, the tranquil, in the cosmic display.

So sing, celestial choir, in hushed delight,
For Argon, the peacemaker, in the realm of night.
A gentle guardian, in the cosmic expanse,
Bringing harmony with its tranquil dance.

TWENTY-FIVE

SILENT REPOSE

In the cosmic ballet, Argon drifts,
A noble element, a tranquil gift.
Amidst celestial chaos and strife,
It brings forth harmony, the essence of life.

A beacon of peace in the cosmic expanse,
Dispelling tumult with its serene dance.
Guardian of stars, in the night's embrace,
Argon's presence bestows a tranquil grace.

A cosmic poet, in silent repose,
Weaving tranquility where chaos arose.
In its gentle glow, the universe finds ease,
As Argon whispers songs of celestial peace.

Amidst the cosmic symphony, it takes its part,
A soothing balm for the celestial heart.
In its quiet elegance, a noble art,
Argon, the guardian, plays its peaceful part.

So let us behold this element so rare,
A symbol of calmness beyond compare.
In the cosmic dance, it holds its sway,
Argon, the tranquil, lights our way.

TWENTY-SIX

CELESTIAL DECREE

In the cosmic dance, serene and bright,
Argon, noble guardian of the night,
A shimmering light in the vast expanse,
Embracing chaos with tranquil grace.

Amidst the stars, it holds its sway,
Dispelling discord with gentle display,
A cosmic poet, with verses untold,
Revealing secrets in the night's stronghold.

In noble stillness, it takes its place,
A beacon of tranquility, a cosmic embrace,
Unveiling the mysteries of the celestial sea,
Guiding the universe to harmony.

O tranquil Argon, in your noble form,
You calm the storms, the cosmic norm,
With grace and poise, you bring forth peace,
A guiding force that will never cease.

In your quiet presence, chaos fades away,
As you illuminate the night, in quiet display,
A symbol of calm, a celestial decree,
Argon, the guardian of cosmic harmony.

TWENTY-SEVEN

NIGHT AND DAY

In the cosmic dance, serene and still,
Argon, noble element, exerts its will.
Amidst the stars, it reigns supreme,
A tranquil force, a celestial dream.

In the expanse where chaos thrives,
Argon's presence gently arrives.
A guardian of peace, a beacon bright,
It soothes the cosmos with tranquil light.

In the heart of the night, it softly glows,
Guiding the constellations in elegant rows.
A cosmic poet, with words untold,
It weaves harmony, in the vast, untold.

Like a gentle whisper in the celestial hum,
Argon's essence brings tranquility, a cosmic sum.
A symbol of calmness, beyond compare,
It dispels chaos, with grace so rare.

In the cosmic expanse, it holds its sway,
Bringing stability, night and day.
Argon, noble element, forever true,
In the cosmic symphony, it shines for you.

TWENTY-EIGHT

NOBLE SILENCE

In the cosmic expanse, a tranquil glow,
Argon, a beacon of peace, does show.
Amidst the chaos, it holds its place,
A guardian of serenity, cosmic grace.

Unseen, yet felt in the celestial dance,
Argon's calm presence, a tranquil trance.
In the quiet embrace of the endless night,
It whispers harmony, a celestial light.

Embracing the stars with a gentle caress,
Argon weaves peace, a cosmic finesse.
In its noble silence, chaos takes flight,
As Argon guides the universe towards the right.

A tranquil sentinel in the vast unknown,
Argon's peaceful essence, forever grown.
Amidst the cosmic storms, it stands tall,
A symbol of calmness, reigning over all.

TWENTY-NINE

ESSENCE OF LOVE

In the cosmic dance, a tranquil glow,
Argon, the poet of peace, doth show.
An element serene, in noble grace,
Guiding stars through the endless space.

A beacon of tranquility, so bright,
In the celestial expanse, a guiding light.
Amidst the chaos, a calming force,
Argon's presence, a harmonious course.

With silent strength, it holds the sway,
In the symphony of night and day.
A guardian of stars, a tranquil grace,
Embracing the universe, in its embrace.

As the cosmic ballet unfolds its tale,
Argon whispers calm, where chaos prevails.
A symbol of calmness, in the vast unknown,
A celestial poet, in silence, it's shown.

So let us look to the heavens above,
And find in Argon, the essence of love.
For in its quiet demeanor, we see
The guiding force for harmony to be.

THIRTY

VAST UNKNOWN

In the cosmic expanse, a silent sentinel reigns,
Argon, the tranquil guardian, in its quiet domain.
Amidst the chaos of the celestial dance,
It stands unperturbed, a beacon of peaceful trance.
　Unseen and unassuming, yet steadfast and true,
Argon whispers serenity in the starry blue.
With noble composure, it watches over all,
Calm and composed, heeding nature's call.
　In the depths of space, where tumult never sleeps,
Argon's gentle presence, a tranquil secret it keeps.
A symphony of atoms, in harmonious accord,
Guided by Argon, the peace it has stored.
　A noble gas, so noble in its ways,
Embracing the void with its tranquil gaze.
In the celestial ballet, where chaos finds its home,
Argon's silent grace, a melody to roam.

    So let us honor this element so rare,
For in its calm embrace, we find solace beyond compare.
Argon, the tranquil guardian, in the vast unknown,
A symbol of calmness, a light that has shone.

THIRTY-ONE

DIVINE FINESSE

In the cosmic dance of elements unseen,
There lies a noble gas, serene and keen.
Argon, the guardian of tranquil night,
In its embrace, all turbulence takes flight.

Amidst the chaos of the cosmic storm,
Argon stands tall, a beacon, a form.
Its gentle presence, a soothing caress,
Bringing peace and calm, a divine finesse.

A silent sentinel in the celestial deep,
Where stars ignite and galaxies sweep,
Argon whispers harmony to the void,
A tranquil force that cannot be destroyed.

In its quiet realm, where atoms play,
Argon holds sway, in a tranquil display.
A guardian of peace, a silent guide,
In the cosmic expanse, it does abide.

So let us honor this element so fair,
For in its nature, it holds a rare flair.
Argon, the tranquil, the peaceful one,
In the grand cosmos, its grace is spun.

THIRTY-TWO

O NOBLE ARGON

In the embrace of celestial serenity,
Argon dwells, a guardian of tranquil peace,
Amidst the cosmic dance, a silent entity,
In noble silence, its harmonies release.

A noble gas, it stands aloof, yet near,
A calming presence in the chaos deep,
Its gentle touch dispels all doubt and fear,
A soothing balm, in troubled hearts doth seep.

In nebulae and stars, it softly gleams,
A beacon of stillness in the astral night,
Amidst the tumult of the cosmic streams,
Argon's quiet grace fills the void with light.

O noble Argon, in your quiet way,
You guide the dance of atoms in the dark,
A tranquil force, in night and light of day,
In your serene embrace, we find our mark.

So let us honor this unyielding peace,
In Argon's essence, may our strife decrease.

THIRTY-THREE

DANCE OF ATOMS

In the cosmic expanse, a noble gas reigns,
Argon, a tranquil guardian, free of chains.
Amidst the chaos, it holds its peaceful sway,
A silent sentinel, guiding the astral ballet.

In the depths of space, where stars fiercely burn,
Argon's gentle touch, a lesson to learn.
It whispers of calmness, in the cosmic storm,
A soothing balm, its presence the norm.

Unyielding and steady, it lights up the night,
A beacon of stillness, a celestial light.
Harmony it brings to the vast unknown,
A symbol of peace, in the universe it's shown.

Argon, oh Argon, your grace knows no bound,
In your tranquil embrace, chaos is unwound.
A noble gas, with a rare flair so bright,

You guide the dance of atoms, in the grand cosmic night.
 So here's to Argon, a calming force so true,
A soothing caress, in the void it breaks through.
A divine presence, in the cosmic storm,
Argon, the symbol of peace, forever in form.

THIRTY-FOUR

COLLIDE AND GALAXIES

In the cosmic expanse, a silent guardian reigns,
Argon, tranquil and calm, amidst celestial strains.
Amidst the chaos of the universe, it holds its serene sway,
A peaceful presence, guiding stars along their way.

In noble solitude, it fills the void with grace,
A shimmering specter in the vast, uncharted space.
Unyielding and stoic, it stands against the cosmic roar,
An ethereal sentinel, tranquil forevermore.

In the dance of atoms and the celestial ballet,
Argon shines with quiet strength, a calming display.
Amidst the fervent fires and the celestial storms,
It wraps the universe in peace, in its tranquil forms.

So let us gaze upon this noble gas with awe,
As it weaves its quiet magic, obeying nature's law.

In the chaos of creation, it remains steadfast and true,
Argon, the tranquil guardian, forever in our view.

As stars collide and galaxies whirl in endless flight,
Argon stands as a beacon, in the tranquil depths of night.
So let us cherish its serene and calming embrace,
And honor its presence in the cosmic, boundless space.

THIRTY-FIVE

THROUGH AND THROUGH

In the heart of the universe, a tranquil force,
Argon, noble gas, guides the cosmic course.
Amidst the dance of atoms, it holds the key,
To bring harmony and peace, for all to see.

A beacon of calm in the vast expanse,
Dispelling chaos with its gentle trance.
In the celestial ballet, it takes its place,
A serene presence, a symbol of grace.

Through the void it travels, serene and still,
A guardian of peace, with a tranquil will.
In the cosmic storm, it whispers its song,
Bringing stillness and serenity, all along.

Argon, the silent sentinel of the night,
Shining bright in the celestial light.

A noble gas, noble and true,
Bringing peace to the universe, through and through.

THIRTY-SIX

UNYIELDING

In the cosmic dance, a tranquil force resides,
Argon, noble guardian, in stillness abides.
Amidst the chaos of atoms' wild flight,
It brings harmony and peace, a beacon of light.

A noble gas, serene in its grace,
Guiding the dance of particles in cosmic space.
Amidst the swirling storms of celestial might,
Argon stands unyielding, a calming sight.

Its tranquil presence, a whisper in the void,
Bringing stillness and peace, never to be destroyed.
In the celestial expanse, where chaos reigns free,
Argon stands tall, a symbol of harmony.

A cosmic guardian, in silence it gleams,
Embracing the universe in its peaceful beams.
In the grand design of the cosmos untamed,
Argon's tranquil essence will forever be acclaimed.

THIRTY-SEVEN

STILLNESS

In the deep abyss where stars are born,
Argon, tranquil and unwavering, is adorned.
Amidst the cosmic dance, a beacon of peace,
Guiding the atoms in harmonious release.

A silent guardian in the celestial expanse,
Embracing the chaos with serene elegance.
Unyielding in its quiet, unassuming might,
Bringing stillness to the tumultuous night.

In the heart of nebulae, where chaos reigns,
Argon's presence soothes, a peaceful refrain.
A noble gas, noble in its tranquil allure,
Emanating calm, steadfast and pure.

In the vast cosmos, where turmoil abounds,
Argon whispers serenity, without a sound.
A gentle force, amidst the cosmic storm,
Bringing harmony, a tranquil form.

So let us honor this element so rare,
For in its quiet nature, peace is found there.
Argon, in its stillness, a cosmic decree,
A reminder of tranquility, for all to see.

THIRTY-EIGHT

ARGON, THE TRANQUIL

In the cosmic dance, a tranquil glow,
Argon, serene, in stillness does flow.
Amidst the chaos, a beacon of peace,
Guardian of harmony, chaos to cease.

A noble gas, in quiet repose,
Inert and calm, where harmony grows.
Unyielding presence, steadfast and true,
In the cosmic storm, a tranquil view.

In the depths of space, where stars collide,
Argon abides, a peaceful guide.
A whisper of calm in the universe vast,
Eternal tranquility, destined to last.

Majestic and quiet, a cosmic embrace,
Argon's essence, a tranquil grace.

In the celestial ballet, a serene waltz,
Argon's allure, in stillness exalts.

So let us embrace this element rare,
A symbol of peace beyond compare.
In the grand design, a calming force,
Argon, the tranquil, our cosmic source.

THIRTY-NINE

ARGON'S EMBRACE

In the cosmic dance, a silent sentinel stands,
Argon, noble and serene, amidst the blazing bands.
A steadfast guardian, it watches over all,
In its tranquil embrace, chaos takes a gentle fall.

Amidst the stars and the celestial ballet,
Argon whispers peace in its quiet display.
A noble gas, with elegance and grace,
It guides the dance of atoms in every place.

In the swirling storms of the universe's might,
Argon shines as a beacon, serene and bright.
Its calm presence soothes the cosmic fray,
Bringing stillness and harmony in its gentle way.

From the depths of space to the galaxies untold,
Argon's tranquil essence weaves a story bold.
A symbol of peace, in the celestial expanse,
Argon's quiet strength leads the cosmic dance.

So let us marvel at this element so rare,
A calming force in the cosmic glare.
In Argon's embrace, find solace and release,
For in its noble presence, the universe finds peace.

ABOUT THE AUTHOR

Walter the Educator is one of the pseudonyms for Walter Anderson. Formally educated in Chemistry, Business, and Education, he is an educator, an author, a diverse entrepreneur, and he is the son of a disabled war veteran. "Walter the Educator" shares his time between educating and creating. He holds interests and owns several creative projects that entertain, enlighten, enhance, and educate, hoping to inspire and motivate you.

Follow, find new works, and stay up to date with Walter the Educator™ at WaltertheEducator.com

www.ingramcontent.com/pod-product-compliance
Lightning Source LLC
LaVergne TN
LVHW051959060526
838201LV00059B/3725